NATURALLY:

Medicinal Herbs and Edible Plants of the Canadian Rockies

by Rick Langshaw

Edited by Thomas Owen

A Summerthought Publication
Box 1420, Banff, Alberta

ISBN 0-919934-13-7

Printed and bound in Canada

Published by
Summerthought Ltd.
P.O. Box 1420
Banff, Alberta T0L 0C0

to my Dad

Introduction

In order to compile a concise document on herbal medicine it would probably require more books than would fit in the back of a one-ton truck. The diversity of plant types, the changing climatic conditions under which all these plants grow, and thus the quality and type of medicinal use to which they can be put, forces an exponential growth in the amount of information forthcoming. When we considered doing a book of this type, we asked many friends just what they would like to see in print. The overwhelming reply was simplicity and the pertinence to the material contained; in other words, "Am I going to find the plants mentioned in this book in the area in which I live?" As we have found in the past there are many books available on herbal medicine, but usually over half the material contained is not applicable, so we have produced in the book, **NATURALLY: MEDICINAL HERBS AND EDIBLE PLANTS OF THE CANADIAN ROCKIES,** some of the most common and effective plants you are sure to find right here in your backyard. By no means have we attempted to treat all or even part of the knowledge available on this subject in this area; for that you will have to study further, possibly from some of the texts mentioned in the bibliography. What we have done here is to give an introduction only. This subject is more than fascinating and we highly recommend further pursuit of accessible knowledge.

Wild plants growing in this area of the world often face very short growing seasons and difficulty existing at the best of times. With this in mind I would like the users of this book to approach the topic with what we would like to call "an etiquette for nature." In short, this would probably translate into an appreciation for the growing plants that you are about to use; an appreciation that they are living things just like you and that life can be hard for them; an appreciation that they too must live to reproduce their own kind and that an unthinking pillage upon these plants will eventually result in their extinction or decimation. There are many ways to approach a reasonable use of these living things while letting them remain for the use and benefit of others. In our opinion this approach is one of moderation; that is, pick only one of every 10 plants and rather than using a whole flower when the petals alone will do, try to concentrate more so on obtaining only what you need. Besides, part of the joy of going and gathering your own herbs is the time spent in doing so! Much to our delight, most of the plants mentioned in this book are often considered to be weeds. They are constantly being persecuted for their uselessness when with a little knowledge they can become sources of great benefit.

For those intending to roam throughout the Rockies in search of the plants herein mentioned, let us remind you that picking wild flowers in the National Parks is illegal. This legality however is often enforced with reason; for example, the picking of wild berries such as raspberries, or the reasonable collection of a few mushrooms may be viewed as a reasonable "usage" of the resources of a Park. It's when someone is caught with a shopping bag full of delicate orchid roots, or when someone foolish enough to string-out a shopping-cart load of dried mushrooms across a picnic site that trouble can be expected. Throughout most of the Rockies the plants mentioned in this book can be found outside the National Parks, so we would like to recommend collecting them where you are sure of the legalities.

The following is a list of some of the etiquette that should be observed when out gathering:

1. Attempt to gather only one out of every 10 plants.
2. Of the plants where you require the leaves, gather only the topmost.
3. Of plants from which you need the roots or rhizome, use a small digging tool like a fork and leave the surrounding area untouched. Also, gather only one out of every 10, leaving some of the root undisturbed in the soil.
4. If it's mushrooms you are after, remember that mushrooms are the fruiting bodies of the living parts under the ground, so leave some remaining from every bunch so as to reproduce more.
5. Don't break branches when collecting berries, bark or leaves from trees or bushes. This opens the plant to infection.
6. Whatever you gather, please be sure to use it.
7. It is often possible to gather, very selectively, things which you need. For example, instead of taking a whole rose flower for its petals, take just the petals and leave the rest of the flower to produce the rose hips or berries later in the year.
8. Gather only small amounts of what you need, so that you are not left with things that need drying or curing that you simply cannot look after. I suggest going out several times, gathering a little at a time.

Arnica

Arnica is one of the most sought after treasures for natural healing. Caution is advised here, for this is strong medicine especially when taken internally.

Habitat

Here in the Rockies we have nearly 20 species. Many are found at or above treeline. When collecting, be wise and take only a few of the many you may see, for this can be a harsh climate for the plant to multiply in. Look for a tall (15 inch) stalk topped by a multi-petaled yellow flower. The leaves are on opposite sides of the stem, in pairs. There may be one top flower with two opposing less mature flowers further down the stalk, depending on the species.

Preparation and Uses

Gather the rhizome and flowers when the plant is just starting to flower. The flowers will generally tend to go to seed while you are drying them, so try to gather them while they are quite immature. Dry quickly in a warm — not hot — dry spot and seal in a jar. The most effective use of the plant's medicine involves the following: put two teaspoons of dried and ground stems, flowers and rhizomes in one cup of brandy or medicinal alcohol. Seal in a jar for about three days. Use this as a tincture or spray.

Another combination is two teaspoons of the dried flowers, stems and rhizome in one cup of boiling water, simmer for 10 minutes and cool.

This liquid solution may be used on **un**broken skin to stop the pain of bruises, swollen feet, etc. For open wounds re-dilute this solution by adding twice the amount of pre-boiled water.

For internal use after any emergency such as broken bones, concussion or swelling, use **no more** than five drops about every four hours in a glass of water. Children require much less.

Alder

It can grow up to five metres in height while remaining very supple and resistant to breakage. Due to this, hiking through an Alder patch may become near to impossible, as the boughs and branches may seemingly try to trip you.

Habitat

Look for Alder on lower avalanche paths and along river banks. The leaves are oval shaped with serrated edges. The fruit is an acorn-like nut, usually green in colour and rough, like sandpaper. They usually grow in groups somewhat like grapes.

Preparation and Uses

Gather the outer and inner bark, drying and aging it for up to several weeks. Unaged bark will provoke vomiting and pains in the stomach. After aging, the bark's effect is similar to that of rhubarb for constipation. For this purpose, mix approximately 1/8 cup of powdered bark to one cup of brown apple-cider. Take one teaspoon three times a day after meals for about three days or until relief is obtained. Alder bark boiled in vinegar is a good remedy for lice and their associated symptoms — approximately 1/8 oz. of bark per cup of vinegar. This also makes an excellent mouth-wash when diluted with tap water. Use a half and half ratio. This bark may also be used for smoking fish and meats, resulting in a unique flavour.

Bearberry

For years I heard about Kinnikinnick being smoked by the Indians and I really didn't know just what it was. After spending some time with a full-breed Indian from the Peigan Reserve, I was happily introduced to the pleasure of smoking the crushed, dried leaves of Bearberry from a real Indian pipe.

Habitat

This herb is readily identifiable in the fall due to its red berries. During the spring and summer one must look for a leathery, waxy-leafed plant growing close to the ground on dry, sunny hillsides. The shiny, evergreen leaves usually grow from a stem with reddish, shreddy bark. Gravelled slopes and open coniferous woods are a good area to start looking.

Preparation and Uses

Bearberry is one of only a few medicinal herbs useful in treating high-sugar blood due to diabetes. As well, many afflictions of the urinary tract are treatable with the following recipes:

Soak the leaves in sufficient alcohol to cover them and let stand in a covered dish for approximately one week. Use one teaspoon of the soaked leaves per cup of boiling water and drink two to three cups per day. The remaining alcohol can be used by placing 10 to 20 drops in ordinary herb tea and drinking a capful two to three times per day.

An ordinary herb tea can be made by placing one teaspoon of dried leaves in a cup of boiling water, steeping 15 minutes.

Chamomile

For quite some time I was unsure of this plant in its Rocky Mountain setting. It looks like a small, yellow-headed daisy without the petals, although most books describe it as having petals. Our particular variety has **no** petals and smells like a mixture between pine and apple. The leaves are feather-like.

Habitat

I have generally found the best place for Chamomile is alongside roadways with very hard packed soil, where it may grow only a few inches tall in a low covering mat. When the soil is more moist and less packed, it may get to be 10 inches or more.

Preparation and Uses

Chamomile tea may be purchased in most food stores for aiding digestion. This is because it acts as a calming agent for nervous stomachs. Women should take note that this herb promotes menstrual flow and aids in calming uterine spasms. Try to gather the buds when they are most abundant. Dry slowly in a cool place and then seal in a jar. One teaspoon of the plant in a cup of boiling water, steeped for 15 to 20 minutes, is most effective. Several cups may be necessary for an adult while two to three teaspoons will suffice for a child. The cooled tea may also be used as a surprisingly good hair rinse.

Chickweed

Gardeners are usually annoyed by this weed yet few know that its properties as both an edible plant and healing herb probably surpass any vegetable they are trying to protect from it.

Habitat

Chickweed is found in cities and in the country, in gardens and in open fields. Look for it in shaded, moist and unused ground. It forms what looks like a matted green carpet in these areas. The flowers are in clusters with white, leaf-like petals.

Preparation and Uses

Try Chickweed cooked in soups, casseroles and other dishes enhanced by a touch of green. It is also quite enjoyable just as a nibbling herb in salads. Nutritionally it is high in copper and iron. Especially good for women.

The healing properties of Chickweed are justifiably famous. For nearly all skin sores and wounds its healing properties are unrivalled. Apply the freshly washed herb directly on the affected area. Best results are achieved by first crushing the leaves. It may also be used dried and powdered in the same manner. Hold this preparation in place with a bandage and replace every few hours.

For serious constipation, take three heaping tablespoons of the fresh (or dried) herb and boil in a quart of water until it boils down to one pint. Drink a warm cupful of this liquid every three to four hours until relief is obtained.

Cinquefoil

This bushy plant is best identified by its bright yellow flowers which bloom most of our summer. Cinquefoil can be found thriving at quite high elevations here in the Rockies. Generally, we have a plant that grows to about one-half-a-metre high. The leaves are finger-like and the flowers are canary yellow with five petals. The bark is brown and shreddy.

Habitat

Look for this plant on calcareous or limestone based soils. This may include river flats, dry meadows or cliffs.

Preparation and Uses

Gather the roots from this plant and dry them for later use. The root, powdered, can then be used in a tea or as a poultice for the following ailments: as a mouth and throat gargle for poor or infected gums, throat and mouth (a little honey mixed with the tea is the most effective use); sores of the skin or hardened areas of skin should recover quickly when a poultice made from the tea is applied and refreshed several times a day. If response to this treatment is slow, try boiling the root in vinegar and then apply this solution to the affected areas.

Caraway

Habitat

It is found in waste areas such as unused garden spaces and along roadsides. Caraway can easily be confused with other herbs such as shepherd's purse until you squeeze a little of the flower or new seed; a very fragrant licorice-like smell is evident. It grows about one metre high in bushy groups. The flowers are small and white.

Preparation and Uses

The most common use is as a flavouring in breads, salads, baking and stews. A tea may also be made from ½ teaspoon of dried seeds in a cup of boiling water. This is very good for digestion and colic in children. Gather the seeds in late September, then dry and bottle them. May we suggest it as a flavouring with other herbs in tea?

Clover: White or Red

The Indians used Clover for both its nutritional and medicinal properties. From a nitrogen-fixing root system a brush-like, multi-branched stem grows upward to approximately ½ metre in height. White or red flowers top the stem.

Habitat

It is found mainly in cultivated fields but spreads easily to roadways. In the Canadian Rockies it is often found alongside roadways below 5,000 feet.

Preparation and Uses

The leaves may be picked and used fresh or dried. If gathered by the roadside, be sure to wash well. Finely chopped leaves are delicious in salads or may be used fresh or dried in soups or cereals. The plant contains large amounts of vitamins A, B, D, E and K as well as iron, phosphorous, chlorine, silicon, magnesium, calcium and trace elements. The seeds, dried, can be sprouted during the winter. The dried leaves, combined with mint leaves, produce a fine tea.
Clover acts to relax a hyper-acidic stomach. An excellent cough remedy is prepared from Clover blossoms, onion juice and honey. Warm these together over low heat for 15 minutes. I recommend one teaspoonful each hour.

Dandelion

Most of the Dandelions found in the Canadian Rockies are the introduced type; the native species are rare and generally confined to the alpine areas. These plants are perennial, contain a milky-white, very bitter juice inside their fleshy tap root and may be identified by a yellow umbrella shaped flower at the end of a hollow tube-like structure. The leaves are coarsely toothed, usually directed downward.

Habitat

Most anywhere there is covered soil with enough water to support life, you will find Dandelions. If you are going to collect these plants anywhere, either in wild areas or your own garden, **be careful!** Many persons try to eradicate Dandelions with poison.

Preparation and Uses

The root is the most important part, although the fresh, young leaves may be used in salads for variety. Collect the root while the plant is in flower. Carefully wash the root without breaking it, or the juices will flow out. The root is then dried slowly in a warm, dry spot away from sunlight. Dandelions contain about 7,000 units of vitamin A per ounce and is an excellent source of vitamins B, C and G. It is very useful for most skin diseases and urinary problems.
The dried root is powdered and may be used as a coffee substitute. It may also be mixed with roasted, ground nuts in the preparation of a very rich, healthy drink. Prepare and use the grounds as you would coffee. **Do not** filter the grounds — steep them for 15 minutes. If you wish to remove the slightly bitter taste from the leaves for use in salads, soak them for about half an hour in salt water, then sprinkle with lemon juice.

Elder or Elderberry

Most everyone has heard of Elderberry wine, jam, jelly or pie. The berries themselves are high in organic iron. The entire plant from the roots to the leaves is medicinal.

Habitat

It is generally found in low, damp grounds, thickets and roadsides. I've never seen it above 400 feet in elevation. When topped with purple-black berries it is easy to identify. It grows from two to four metres high, blooming in July with small star-shaped flowers. The leaves are large, usually five to a bunch. The odour is aromatic and sweet, becoming sickly sweet after a frost. The large clusters of berries should come off the holding branch as a clump if they are ripe. **Do not** use if unripe.

Preparation and Uses

The Elderberry contains 90% seed and 10% juice, so I usually make jam instead of jelly to avoid wastage. Dried leaves, berries, bark and root can all be used for the following ailments: headaches due to colds, rheumatism, anaemia, complexion aids, swelling of the joints, sore throats and constipation. Be careful when using the inner bark as it can cause vomiting in large doses.

Fireweed

This is probably one of the most abundant plants we have growing in the Rockies. The magenta to pink flowers are seemingly everywhere. Towards the end of August and late into the fall, frost turns this plant the most exquisite shades of red and orange. After the fall frosts or late in the season, puffy silken bundles adorn the plant and fill the air downwind.

Habitat

This plant thrives in areas recently burned, hence the name Fireweed. It can also be found just about anywhere there has been human impact like old logging sites, along roadsides or disturbed areas. One of the areas we look for it is along wide river plains.

Preparation and Uses

When collected strictly as an edible vegetable it should be gathered just shortly after it comes through the ground, until it reaches about eight centimetres. At this point its flavour rivals the finest asparagus. Try steaming it for best flavour. After the plant gets much taller it can be quite woody.

The medicinal qualities of Fireweed come from its astringent properties, or the ability to arrest discharges. Any symptoms occurring from the gathering of mucous can usually be treated with a tea made from the fresh or dried plant. One teaspoon of the leaves in a cup of boiling water and steeped for one half hour will relieve pain due to colds, allergies, hay fever or tonsilitis. Take the tea hot and in small amounts until relief is obtained.

Horsetail

This plant is one of the most ancient of land-living plant forms. The stalks are segmented and may be pulled apart in sections. Eight species are known within this area and we have only tried one *(Equisetum Arvense L.)*. This type seems to be the most common. Its identity is usually associated with a bush-like growth showing much branching. Other varieties tend to have one main stalk with fewer branches. It is a short plant, 10 to 15 centimetres high.

Habitat

Found growing in moist, generally shaded areas under trees or higher bushes and grasses, the plant contains much silica, which adds to its medicinal quality.

Preparation and Uses

Preparation of Horsetail is used for eye and skin treatments because of its silica content. Skin disorders and offensive perspiration such as feet odour respond well to the following solution: steep one teaspoon of dried Horsetail in one cup of boiling water for 45 minutes in a covered container. Cool and use as a wash. For relief of discomfort and difficulty of discharging urine, use one mouthful of the above tea, four times a day. Bleeding of the stomach or intestinal track may respond well to this tea also.

Lodgepole Pine

The long, straight stands of Pine seen in the Rockies are the Lodgepole Pine. They can reach a height of 25 metres with barely a waver of uniformity.

Habitat

Generally Lodgepole Pine is found below 6,000 feet, in fairly dry terrain.

Preparation and Uses

The most important quality of this Pine is in the use of a resin for treating colds and their associated ailments. For sore throats the gum is chewed. It may also be used as a powder and applied to the throat with a swab. Mix one teaspoon of the powder resin with five crushed Juniper berries and one cup of boiling water. This is an effective remedy for colds and influenza. The heated resin is also useful in drawing out splinters and opening boils or inflamed insect bites.

Juniper

Both of the common varieties are abundant in the Canadian Rockies. One is a bush approximately one metre high and the other is a creeping sort of low bush. Both have bluish-green to blue-black berries. Juniper is a needle bearing plant that stays green throughout the year. It has a sweet fragrant scent often used in sachets.

Habitat

Look for Juniper on dry rocky slopes, ridges and floodplanes. Sunny, wind exposed slopes seem to be a favourite spot. The berries take two years to ripen, which is why one bush may have both blue-green and blue-black berries.

Preparation and Uses

I have found the best time for Juniper berries is after they have turned from blue-green to blue-black. They acquire a much sweeter flavour. The berries are gathered, lightly dried and stored in sealed jars. As time passes, they do not become brittle or hard if stored this way.

Juniper is well known for its medicinal uses as a disinfectant and weak-stomach aid. Stomach pains should be treated by chewing five berries a day for several days then gradually increasing the dosage over a week to 15 berries per day. Then decrease this by one berry a day for five more days. This process may be achieved similarly by making tea with the same amount of berries per day as described above. Try two cups of boiling water over the crushed berries, steeping for at least 15 minutes. Drink all the tea made. If pain persists consult a doctor.

As a disinfectant make a tincture of about one cup of alcohol to 15 crushed berries. Let stand for 24 hours and spray any rooms in which patients with infectious diseases have been staying. As a gargle use the above mixture half diluted with water. **Do not** use Juniper in large amounts to begin with or if the kidneys are inflamed.

As a flavouring I can recommend some of the many ideas I have tried:

1. Add about 10 halved berries to vegetable and meat stews.
2. Add about five halved berries to any tomato sauce.
3. Place in jars in which you have stored dried apples. About five berries per jar.
4. Finely crush two to three berries and sprinkle over vegetable dishes such as turnips, cabbage or broccoli.
5. Try a few berries in porridge for a nice change.

Labrador Tea

This was probably the finest herb I ever collected for medicinal use, since it is easily found and makes for a very pleasant tea.

Habitat

Look for this plant near or beneath dry Lodgepole Pine. You can immediately recognize the plant by the soft brown underside of the leaf and by the narrow, long, green upper leaves. They turn lovely shades of orange as autumn approaches.

Preparation and Uses

I suggest caution in the gathering, drying and use of this plant; it contains a very active oil of the name Ledum. This oil can affect the heart if taken in large doses. Ensure ventilation when drying this plant.

I could probably recommend no better remedy for coughs, colds or bronchial affections.

A tea made with one teaspoon of dried leaves per cup of boiling water and steeped for 15 minutes offers best results. For internal use take one cup every half hour, or externally in a poultice — for burns, boils, bruises, insect bites.

Oregon Grape

One of the most common names for this low growing shrub is ''Holly-leaved Barberry.'' When looking for this plant, keep in mind its holly-like leaf which is often shiny and grading in colour from dark green to red. If it was only for the fantastic jelly-making qualities of this berry, this plant would be justly famous. However, some of the medicinal uses of the berry are recently becoming known.

Habitat

I have only found this plant growing in warmer, sunny areas west of the Great Divide. I have heard of its presence in the Eastern Rockies but have yet to find it there. Sunny, dry slopes in open grades of trees are the best places to search. The yellow flowers help to identify it during flowering periods but afterwards it is very difficult to recognize. It seems to fruit but every other year as some years I cannot find a single plant bearing fruit.

Preparation and Uses

I gather the berry for the most to make jelly. The berry is a dark blue and extremely sour. Once gathered, the berries should be used immediately. The most incredible purple-blue jelly is your reward.

The roots may be used as a tonic for ailing persons and give good results as a laxative. If you are prone to skin diseases such as eczema or psoriosis, take two or three teaspoons of jelly per day.

Birch

There are basically two types of Birch present in the Rockies: the standard Birchbark Tree and the Dwarf Birch. The Birch Tree can reach heights of 20 metres or more and is recognized by the white, curling bark. Dwarf Birch is more of a shrub with reddish bark and is very seldom more than three to four metres tall.

Habitat

The Birch Tree is found intermittently in the Western Rockies, usually at elevations below 4,500 feet. Look for it in stands of deciduous trees. The Dwarf Birch can be found on avalanche slide-paths and in most valleys.

Preparation and Uses

Both the leaves and bark can be dried and stored. Tea made from one teaspoon of the leaves or bark in a cup of boiling water, steeped for 15 minutes, is an excellent remedy for diarrhoea. This solution may also be applied externally for boils and skin eruptions.

Plantain

There are two types of Plantain; broad and narrow leaved. Both types are useful. The leaves radiate from the base of the plant, are dark in colouring and are strongly ribbed lengthwise. The flowering stem can attain a length of almost 1½ feet. This is very helpful when trying to identify the plant.

Habitat

Plantain is found along roadsides and sidewalks, often in very hard trodden earth.

Preparation and Uses

The whole plant can either be dried and used later, prepared fresh in a tea, or pounded into a paste. As a tea, use one teaspoon to one cup of boiling water. Let stand for 15 minutes. For relief of insect bites, snakebites or stinging plants, bathe the affected area with cooled tea every hour or apply a paste made from beaten leaves. The most useful quality of Plantain is its ability to soothe and promote healing of sores. Apply the paste directly to new or old wounds. Plantain is rich in vitamins C and K and many essential minerals. The young leaves are a pleasant addition to salads or may be steamed with a dash of lemon juice and butter for a nice change from everyday vegetables.

Rhubarb

This is generally considered to be a domestic plant and often its medicinal properties are neglected. Both the green and red stemmed are found in most gardens. The red stemmed variety is recommended for its milder taste. The plant is crowned by large to monstrous green fan-shaped leaves hiding the stems which you seek.

Habitat

Avoid wild Rhubarb as its medicinal properties can be exceedingly stronger than the domestic variety. This plant breaks through the ground very early in the spring looking like "pink eggs." It prefers rich, well-tilled soil but may be found in most gardens regardless of the ground-type.

Preparation and Uses

Stewed Rhubarb is one of the most pleasant forms of the prepared stems. Simply chop off the leaves and section the stem into one inch pieces. Place these in a **stainless steel** or **porcelain** pot, adding one cup of water for every 10 cups of Rhubarb. **Do not** use aluminum as the active ingredient in Rhubarb will dissolve it. Now add a dash of salt and approximately three cups of sugar. Bring to boil — let simmer for about half an hour. The resulting stew is not only good to eat but is an excellent laxative. Berries, such as raspberries, may be added to make an excellent pie filling. If the stems do not provide relief from constipation, the roots have a much stronger active ingredient, but caution is advised. Prepare a stew as you would with the stems but consume only half a cup per day. If this does not help, consult your doctor.

Sage

One of the most popular of herbal flavourings, Sage is used with chicken, fish and stews. There are a great many plants covered under this name, all of which resemble the garden variety in smell and taste. The majority of plants bearing this name have grey to green leaves which have a rumpled surface. Upon biting a leaf one will notice a warm, bitter taste, somewhat camphorous.

Habitat

On the west slopes of the Rockies, in low valleys and dry, rocky spots are the places to start looking for this plant. There may be a variety of Sage living on the east slopes of the Rockies but I have yet to confirm this.

Preparation and Uses

A tea made of dried leaves is unsurpassed in relieving flatuence (gas pains). One teaspoon of dried leaves per cup of boiling water then steeped 15 minutes is the recommended dosage. This may also be used as a gargle when combined with cider, vinegar and honey.

Shepherd's Purse

This is another of the garden weeds people try desperately to destroy, not knowing what a valuable herb it is. The seeds, which the plant produces in abundance (up to 64,000 per plant), are very high in body building elements and extremely nutritious. Shepherd's Purse contains large amounts of vitamins A and C, calcium, sulphur and sodium. One of its most medicinal properties is vitamin K — the blood clotting vitamin.

Habitat

Most gardens, unused spaces and roadways below 5,000 feet in altitude nuture this herb. Look for dandelion-like leaves at the base with one to several stalks growing two to three feet tall. The plant may be quite bushy and can be identified by the seeds or Shepherd's Purse. The taste is like cabbage and the odour unpleasant.

Preparation and Uses

Collect the seed pods both for drying and immediate use in salads. The young leaves have a pleasant peppery taste and may also be used in salads or steamed. Due to its blood-clotting agent, Shepherd's Purse is recommended for hemorrhages of all kinds. Take internally as a tea or apply externally as an ointment or poultice. One teaspoon of ground leaves or seeds per cup of boiling water — steep for 15 minutes. Bathe wounds with this cooled solution. Hemorrhoids also respond well to this solution in suppository form.

Self Heal

Some common names for this plant are "Heal All" and "Wound Wort." It has a pretty purple flower which very much intrigues me.

Habitat

It is usually found in very moist, shaded places; almost swamp-like. The plant grows in groups and may be readily identified only when flowering. It is most common west of the Great Divide.

Preparation and Uses

I collect the young flowers and plant tops before they have a chance to seed.

Common uses are as a tea (one teaspoon per cup of boiling water — steep 15 minutes) for expelling worms. For external use it will stop bleeding and promotes the healing of mouth ulcers.

Mint

This plant gives us some of our most commonly used flavourings — menthol, spearmint and peppermint. All Mints have square stems surmounted by white, pink or purple flowers.

Habitat

Mint is usually found in moist, cool and shaded areas. I find river banks and spongy ground near ponds the best places to start looking.

Preparation and Uses

Once I've located a Mint patch, I gather only the upper most leaves, leaving enough of the plant to flower and produce seed. Dry the leaves immediately in a warm, dry place and seal in jars. The leaves are usually used as a tea. One teaspoon of dried leaves per cup of water. Mint is also an excellent flavouring with lamb, peas or vegetable stews. Besides having a pleasant taste, Mint is effective in soothing indigestion, headaches and toothaches.

Sorrel

There are a number of members to this plant family; the one found in the Rockies is common in alpine and sub-alpine zones. The plant may be easily identified by the heart-shaped, ground-hugging leaves, and in autumn by the tall reddish flowers. Tasting a leaf will confirm your recognition, should you discover a tart-lemon flavour.

Habitat

Found generally in upper sub-alpine or lower alpine zones of the Rockies, they require a fair amount of moisture and seem to prefer partial sunlight.

Preparation and Uses

Collect and dry the leaves for later use or use fresh, chopped leaves to spice up a salad.

A tea from dried leaves (one teaspoon per cup) will expell worms. Besides aiding a nervous stomach, it is also an excellent thirst quencher.

Valerian

Unknown to most people, this plant was probably the forerunner of our modern day Valium. Considering the cost of this prescription, and its abuse, I highly recommend — though with caution — this herb as a substitute.

Habitat

Valerian may be found throughout the sub-alpine and alpine zones of the Canadian Rockies. It likes damp, partially sunny areas and is quickly identified by its tall stalk topped with a semi-globular cluster of white to pink flowers. The plant tends to be quite aromatic, more noticeably in the autumn after the first frost, when the smell becomes sickeningly sweet. I am told this smell can drive cats wild — something like catnip.

Preparation and Uses

The root of this herb is what is most sought after during its flowering period. At times it becomes quite difficult to dislodge the rambling tube-like root from the surrounding entanglement, allowing only a small percentage of the actual root to be taken. This root — again quite aromatic — should be dried and stored in a sealed jar. Be careful with your introduction to this plant as it can be quite a sedative. Try one teaspoon of the dried, ground root mixed with one teaspoon of dried peppermint in two cups of boiling water. Let stand for at least half an hour. Drink one wineglass full for a calming effect. Wait at least a full hour before trying any further dosage.

Violet

This plant is found in many colour combinations from violet to yellow. There appears to be no difference in medicinal qualities with regard to colouration.

Habitat

Violets flower in very early spring and may continue to flower throughout the season. They are found in moist, shaded areas, preferring humus soil. Violets may grow in alpine zones or as low as sea-level.

Preparation and Uses

More information is becoming available on the uses of Violets as a cancer treatment. A handful of fresh or dried Violet flowers and leaves in a pint of boiling water, then steeped for two hours, is the recommended preparation. Once strained, this solution is applied to the affected surface in a poultice form. The solution may also be taken as a tea for most stomach and bowel complaints. One cup every two to three hours. Consumption of the tea relieves pain from most cancers when taken regularly. For a fever, its cooling properties are renowned.

Wild Rose

During the seasons of late spring and autumn, the Wild Rose offers two completely different treasures for gatherers. The petals of the flowers can be gathered and dried in the spring and the "hips" or berries in the fall.

Habitat

Although the Wild Rose is found throughout the Rockies, it is best looked for in valley bottoms and in sunny areas where it produces more flowers and berries. The thorns on the Rose can be minute to large, so wear gloves when gathering.

Preparation and Uses

Petals: When gathering petals from the flowers, do not pick the whole flower for a berry will not be produced. The petals should literally fall off the flower when gently shaken. Immediately spread the petals on a meshed screen and dry in a warm, dark place. Do not sun or oven dry as they will loose their subtle scent. Once dried to a point where they will easily crumble, seal them in airtight jars. When opened months later in winter, the wondrous fragrance of Wild Roses will fill your room. Use them in tea for a gentle flavour, or in sachets.

Rose Hips: Undoubtedly the most outstanding reason for gathering the Rose hips is its high content of vitamin C. I have mentioned that, unit for unit, Rose hips contain 25 times more vitamin C than an orange.

Heat destroys vitamin C, so preparing jellies and jams is not the most efficient use of Rose hips. The best method is to gather the hips as they become red and dry them in a warm, dry spot. These dried hips are then ground and may be added to desserts or already cooked foods. The ground hips may also be used as a sugar substitute for people on low sugar diets.

When preparing the hips as a jelly, cook them as little as possible. They tend to have a very mild flavour, so I usually add other berries or fruit to enhance the taste. Cranberries and apple, when added together, make a lovely jelly.

Willow

The latin name for Willow contains the word "salix," from which is derived our common aspirin: acetysalicylic acid. Long before aspirins were ever conceived, Indians were chewing the bark of the Willow to relieve toothaches, headaches, fevers, arthritis and colds.

Habitat

Over 300 species of Willow grow in varying sizes from about one inch to 100 feet. It is difficult to pin-point a specific location for this plant as it grows in so many areas. However, I do recommend looking in places surrounding bogs, lakes or rivers. It grows very well in wet places and in spring may be identified by its flower or "pussywillow."

Preparation and Uses

The most effective utilization of the Willow is to boil one ounce of dried or fresh bark in one pint of water, let steep for 15 minutes, then take one wineglass-full three or four times a day. It may be used whenever aspirin is recommended. "Take two glasses of Willow bark tea and call me in the morning."

Yarrow

The latin name for the Yarrow – *Achilleu Millefalium* – supposedly comes from Achilles, who discovered the medicinal qualities of the plant. The stalks, stripped of leaves and flowers, have been used by the Chinese for centuries in the mystical practice of fortune telling with the "I Ching."

Habitat

Yarrow is a perennial plant found growing in pastures, along roadsides and in meadows throughout the Rockies. It generally grows less than one metre high and is topped by a flat cluster of small white flowers. Its strong sage-like smell will usually confirm this plant's identity. Chew a small piece of an upper leaf if you are unsure; it should taste like a rough astringent.

Preparation and Uses

It is difficult to single out the most important medicinal uses of Yarrow. Almost any respiratory tract ailment such as colds or influenza respond to this herb. It is also a great aid to digestive problems, and the juice, applied to the eyes, will releive soreness. Most irregularities of the bladder also respond well to Yarrow. Personally, I have found Yarrow tea to be not only a stimulant when I'm tired but also to help settle a nervous stomach. Yarrow, combined with chamomile in a tea-form, make an excellent hair rinse when cooled.

Gather Yarrow leaves and flowers by stripping a length of the stalk. This should be dried slowly in a cool, dry place and then sealed in a jar. Mix one teaspoon of Yarrow per cup of boiling water and let stand for 10 minutes. The taste of this plant is very strong so I doubt much can be consumed at once, but try three or four cups a day for relief of any of the symptoms mentioned.

Mushrooms

A walk into the woods anywhere in the Rockies will usually leave a would-be mushroom hunter very confused. There are literally thousands of species of mushrooms available to you, but which ones are edible? Even with an excellent guidebook in your pocket, this is a difficult question to answer. So what do you do? The best suggestion we can make is this: choose two or three varieties which you are sure of and that are plentiful and use these on a regular basis leaving the unknown ones until you are sure of them. So we present here three very identifiable and plentiful mushrooms that are sure to please your palate. Along with these three species are also some suggestions as how to prepare or store them. Collecting mushrooms is viewed by some as hazardous to crazy, but with the following guidelines, which are to be followed meticulously, you have nothing to fear.

Firstly, gather only one type of mushroom at a time; never mix them.

Secondly, make sure that each one of the identifying features is met.

Thirdly, gather only fresh specimens, never old ones and never the young buttons that are not readily identifiable.

Fourthly, use or preserve mushrooms as soon as possible after gathering to preserve freshness.

Fifthly, when trying a new type of mushroom, even a type you are absolutely sure of, try only a little the first time. The reason for this is that some people have allergies to even the most common mushrooms.

We would like to give just a few tips on the storing and preserving of mushrooms. Fungi like mushrooms can be very difficult to preserve and even when done with the greatest of care may not turn out just as you would like. Be gentle and take your time with them. When cooking, cook them slowly over low heat. In a like manner for drying them, be slow and methodical. Use low heat and make sure they are very well dried or they will decay easily. There is one exception to this for the mushrooms we recommend but that will be dealt with for that particular fungi.

Lactarius Deliciosus (Delicious Lactarius)

The cap is carrot orange and with age and bruising becomes greenish, grey-green or orange-green. The middle of the cap is usually indented to form a cup. The edges of the cap become ruffled or indented as the fungi grows. The **most important** identifying feature of this mushroom is that it lactates or bleeds orange when broken or cut. Often the cap is slow to bleed but a knife cut across the stem should **always** produce a bright orange milk. Bruising to green does not lessen the quality of the flesh although it does not look pleasant. This mushroom very quickly is infested by little white worms that burrow up through the stem. You will probably find it quite discouraging when gathering this species as it is hard to find without worms. The worms do not affect the taste or quality but do affect your desire to eat it. Try gathering the younger bodies that have not reached full size but ones that are no longer in the button stage.

Habitat

This species tends to form a type of relationship with certain pine and spruce trees, with whom a sort of parasitism is formed with the roots. So look for this species in moist but well drained mossy soils surrounding spruce or pine trees. It fruits in late summer and late into the fall. It can be found long after the first frosts, which tend to turn its colour to green or grey -green.

Preparation

This mushroom has an excellent flavour both fresh and after drying. It should be diced into half inch slices and cooked slowly over low heat if eaten fresh. If you plan on drying this species we recommend slicing it very finely, no more than 1/8 inch thickness and then drying it slowly at a temperature of about 90 degrees Celsius. When crunchy-dry, seal it in an airtight jar. It can then be reconstituted in soups, stews or with fresh vegetables.

Shaggy Mane and Inky Cap

Technically these are two different species but both bear so many of the same attributes and qualities that we have considered them together. They are both members of the *Coprinus* family of mushrooms and are both excellent eating. These are fall fruiting fungi and during this time can be found by the hundreds, together as groups or spaced out in long lines only a few inches apart. The fruiting body may stand up to 20 centimetres high with the cap dropping down over the stem to form an umbrella-shaped structure. The surface of the cap is always shaggy like the hair of an English sheep dog. This surface tends to gather gravel and sand so be sure to wash them off or try gathering them when it is raining or very damp. The **most important identifying feature** of these mushrooms is their ability to self-digest themselves as they become older. This digestion appears as a black ink that drips from the edge of the cap until the entire cap has dissolved. This happens quite quickly so the mushrooms must be gathered and used or dried before this process begins to occur. Look inside the cap of all specimens and if the inner part of the cap has a leading edge which has become black or pink rather than a buff white, it's probably unusable although not poisonous. These fungi have an unbelievable pressure behind them when growing and can lift pavement and sidewalks, cracking them in the process.

Habitat

Look for these mushrooms in large groups just about anywhere there is or has been buried wood. The best spots seem to be along roadways, lawns, old buildings, fireroads and old mill sites. It is not unreasonable to expect to find several hundred of these fungi together in various stages of maturity.

Preparation

These mushrooms **must** be gathered and prepared before it begins to turn black or you will be left with nothing but black goo. When cooking, do so at a low heat for about 10 minutes. Once the gathered portion has been heated above the boiling point for a few minutes, the enzyme which destroys the mushroom is killed and then it will no longer tend to turn to black liquid. This is very important, for if you wish to freeze or dry this fungi for storage, it must be heated to above 100 degrees Celsius and then dried or frozen in the normal manner. Copious amounts of water are given off when drying and several pounds of gathered mushrooms may result in just a handful of dried chips. However, this drying process is very worthwhile for this is the tastiest mushroom we can recommend to you when dried. **Do not** dry at high heats; that is, not over 90 degrees Celsius and do not over-dry as they become very bitter if you are not careful. When drying check them often, as you want them at a stage somewhere between being leathery and crisp.

Oyster Mushroom

This fungi gets its name not from its taste, but from its tendency to look like a white oyster. This is probably the **most important identifying feature** of this mushroom. To be sure with this one, take a spore print from it. This is accomplished by placing the cap, minus the stem, on a white piece of paper with the gill side down. Leave in this position overnight in a quiet spot. In the morning lift the mushroom off the paper carefully and then hold the paper to a strong light. You should observe a beautiful pattern of radiating spirals of a lilac-grey colour after it is well dried. This results from the spores (like seeds) being deposited on the paper. The colour **must** be as mentioned. This is the one infallable method for determining a mushroom's true identity. Good mushroom handbooks will always give spore print design and colours.

The stem should be very tough or woody and the fruiting bodies should always occur together in numbers. The gills are very white and generally flow right into the stem with no break.

Habitat

Look for this mushroom in the fall, growing in areas where there are rotting logs, either exposed or buried. It is often possible to find them growing right from the log itself. Fireroads are excellent places to look, and when you find a few you will usually find a great deal more.

Preparation

This mushroom needs no special attention. When cooking observe the rule of slow cooking. Also, this fungi should be sliced quite thinly if you are going to dry it. Never dry any mushroom in the sunlight as this will detract from its flavour. Dry it at a very low heat and then store in a sealed jar. As with the other mushrooms we have mentioned, once dried it can be used in a great variety of ways. They may even be powdered once dried and used as a sprinkle seasoning.

How to Make a Sachet

Materials needed:

1. A piece of cotton fabric 5″ by 13″.
2. A piece of quilt-stuffing material the same size as the cotton. This is used to wrap the dried herbs in so that they will not filter out of the cotton as they become crushed finer due to handling.
3. A piece of ribbon 6″ long.
4. Dried herbs.

Assembly instructions:

1. Turn the 5″ edges down ½″ and sew.
2. Fold the cotton in half and sew ½″ seams up the sides.
3. Wrap the dried herbs in the quilt-stuffing material so that it cannot leak out.
4. Stuff into the bag sewn.
5. Tie closed with the ribbon.
6. Seal in a plastic bag until used, to retain the smell.

When storing dried herbs to be used in sachets, place a drop of cognac in the jar to preserve the scent. A few of the combinations I have tried are apple, rose petals and basil; or juniper, chamomile and thyme. These sachets make lovely gifts and help keep your sheets and towels smelling nicely all year long.

Index

splinters, 21